Kris Meier

My car has to go!

One way or another

One way....

She takes a swift to work every work day at this specific time so she doesn't have to order an extra one today. Punctually as always the van including nine seats comes humming up to her door. After getting in through the generously wide side door, she takes her seat in her "office space". This sound-proof space is set apart from the rest of the passengers area. Although it is slightly more expensive than an "open-space" seating, it is a great possibility for her to have the 39 to 42 extra minutes which the trip to her workplace takes to focus and prepare herself for the day.

This morning she leaves the table folded up, tilts her chair back and relaxes. She would like to arrive to her workplace as fresh as possible as she has a meeting first thing in the morning with representatives of a private firm. She works in the public energy sector and her team is currently busy with the optimisation of human energy, the fourth major source of energy in addition to wind, solar and water energy. The energy sources of the past don't exist anymore, although unfortunately their destructive, life-threatening after-effects are still with us. But that is not on her agenda today. Her task today is to evaluate whether the module developed by the private firm will actually improve the storage of human energy on the grid.

Outside there is the humming of lots of 9-seater vans, also some vans with fewer seats and even a few with more. Moving among them are numerous bicycles, pedelecs and even runners.

Some of the runners are simply going quickly from A to B, but many of them are pros carrying letters and small packages for their clients. In the past people would use their own cars to cover these short distances within their own district. For hours they would drive the heavy vehicles often only a few kilometres to take part in running tournaments, where they would run races and then drive back home again. These events were sometimes very big with thousands of participants and long queues to get into the car-parks. Running has always been popular. Now it's finally a recognised, well-paid and very attractive

job. Just like things were before the age of motorised transport. At that time rich people had their own runners, a worthy occupation. There were even relay runners, who carried the news over the Alps. Back then this was the fastest trans-alpine news service.
Of course, the news is no longer transported like this, but letters and small packages are. Delivery times are astonishingly short, and since cities are now free of exhaust fumes, being a runner is looked on as a very healthy profession. For most of them it's not their main occupation, they have managed to turn their hobby into a part-time job. If they wish they can have their times and distances digitally logged. There are official national and international rankings, from which firms that are willing to pay for the best recruit their runners.

It's the morning rush hour. But there is no comparison between this and the traffic chaos there used to be with private cars. She looks out of the window and thinks back to the time when mammoth traffic-jams and horrendous accidents happened on a daily bases, when driving was pure stress, full of aggression and brutal recklessness, when the individual vehicles looked more like weapons of war than means of transport.

How much better things are now. From wherever she happens to be she can order a swift. Within 10 minutes at the very latest one of these communal vans will arrive and take her more or less directly wherever she wants to go. Inside one of these there are very clear rules like regarding polite behaviour. Previously the neglect of common courtesy on public transport had become so far gone that it was only used by those who had no other option. With the changeover to obligatory public transport all inappropriate behaviour was affectively put under wraps. Troublesome repeat offenders have to travel without the aid of a motor, at first for a certain period, but in serious cases permanently. The right to use the swift again can be acquired by attending a course on behaviour on public transport. The bar to complete this course is set very high.

At first the vans had drivers but they became self-driven later on. Every booking, whether spontaneous or planned, is logged in the central control computer, CCC for short. This machine plots a course as efficient as possible for every individual van. Timetables and bus stops that crowded the streets with huge, stinking bendy buses belong to a past now. For all their hugeness they often were nearly empty and followed fixed routes with long waiting times. Since today there are neither traffic-jams nor traffic-lights the average journey time has shrunk to a minimum. Accidents happen very seldom, and when they do injuries are minimal. The exception here are careless or suicidal pedestrians. Sometimes cyclists also sustain injuries but the roads are much safer for them now than in the fossil fuel age, when what was literally a war of each against all was raging in the streets.

For many the abolition of the private car was a challenge. For most of the population a huge part of their lives was devoted in one way or another to their car - rustling up the funds for it, making sure it had all the right accessories and special features, and tirelessly attending to its care and maintenance. When they spoke of living costs back then they were generally referring to their cars and their children and it was never really clear which they were more attached to. Cars spent by far the greater part of their time sitting unused in little rooms built especially to house them. These had automatic doors, and believe it or not, some of them were even heated. By the usage of asphalt made from crude oil, vast areas were turned into tarmac deserts, just to have parking places.

Parks with trees, on the other hand, were few and far between. Inner cities where enslaved to columns of metal buildings surrounded by the stinking of exhaust fumes. Highly efficient air filters prevented individual car users from smelling the stinker in front of them, and made sure they would be the last to realise they were fouling the air just as much. Gigantic turbines shovelled poisonous gases out of tunnels. Air conditioners prevented citizens from having contact with an environment grown unbearable. By the end of the era, indeed, it was no longer possible to open car windows. Hermetically sealed, like aliens on their own planet, these car occupants- usually one alone- rolled through an increasingly barren landscape.

As people became more and more fed up with the state of things on the roads- the endless, daily traffic-jams, the ever-increasing dangerousness of driving- they sought to avoid the situation and following their already insane logic, took the traffic to the sky in so-called flying taxis. Thus the noise level that had formerly been confined to the areas around airports spread everywhere. Of course these airborne vehicles which only the very rich could afford to use, guzzled much more energy than those on the ground, in spite of being hydrogen-driven. It was the ultimate excess in personal mobility.
About the true nature of cars the general mass of consumers were kept in the dark. It may have been a marvel of technological speciali-sation but they were fed disinformation about the destruction involved in its pro-duction. Did anyone know that to make sure

the tyres where properly black, soot was mixed into the rubber, and to get this specific soot tons upon tons of crude oil had to be burnt? Who was told that every year tens of thousands of new tyres were used to heat cement works - just to safeguard the market price? Knowing things like this could have spoilt all the fun of driving, and no one dared to do so. We have since documented in full historical detail what an infernal creation was underway at that time - a progress full of noise, dirt, waste and drudgery. The only part of it, consumers were allowed to witness, was through guided tours of spic-and-span assembly plans, where robots, aided by a few salaried worker with only light work to do, fitted the components together according to strict quality criteria components that had largely been prefabricated in under-developed countries under terrible working conditions. For most people their first experience with these luxury machines was in glossy brochures or in air-conditioned showrooms. There they stood on gleaming tiles, as if they had fallen from heaven and come gently to rest. Smart salespeople patiently and graciously explained everything the newest “wonder model” could do. Smiling ecstatically with a positively galactic face they would present incredible functional features, like head-up displays and things otherwise only found in fighter-jets. So called auxiliary systems, glowingly described, gave prospective drivers the impression they would be safe even in the physical limit zone. And yet everywhere every day there were appalling accidents, where the aftermath looked like a battlefield or a terrorist attack. Since the dead, the injured and the wreckage were all cleaned up very quickly, a radio-announcer would be able to say, in a completely neutral voice: *due to an accident in such and such a place there are likely to be delays.*

Year in, year out, to these sometimes not very intelligently guided missiles, each weighing several tons, up to thousand human lives were sacrificed world-wide. Hundreds of thousands of survivors ended up crippled. In many places there were various kinds of warning signs. But signs with official warnings about previous accidents were not among them. People were encouraged to assume that it always happens to somebody else. Penalties for causing an accident were comparatively light. Traffic surveillance was only sporadic. No politician at that time was keen to annoy his potential voters by bringing in tighter controls. Especially not on a local level, which is why sensitive spots, such as in front of kindergardens or schools, were never really subject to traffic control. Through technology that was partly illegal, but certainly tolerated, most people had a warning system inside their car that always alerted them in time anyways.

In Germany there were no controls on old people with physical or mental impairments, so they were getting unrestricted permission to drive. Many doctors were perfectly aware that many of their older patients should have their driving licence revoked, but they either kept quiet or dished out permits as favours. Words circulated fast if a doctor had the licence removed from someone with dementia. He would soon find himself losing patients. On the one hand the attitude to the cars themselves was very different. They regularly had to go for rigorous technical checks. On the other hand if a driver killed someone on the road he usually got off with a suspended sentence.

Only when the car started to come under increasing social and political criticism did the justice system begin punishing traffic crimes appropriately. Whereas it had already been accepted in almost all countries that even on freeways there should be a speed limit, Germany - highly developed in other aspects - firmly dug its heels in on this point for a long time. It was the biggest producer of cars in the world and the car industry lobby - contrary to the interests of the people - had the say in politics. This no-speed-limit madness was tolerated so that the world would see their products as racing cars, which was good for the number of exports. For this the few dozen extra road deaths and several hundred seriously injured were felt to be a price worth paying. And all the time the majority of people were in favour of speed-limits, especially those who had to be on the road with their children. In this market-dominated rather than industrial period, there was little to be gained politically from public safety measures. Pronouncements made in this regard were never more than lip-service or electioneering.

Citizens were protected, not as human beings, but as tax-paying consumers, which had extremely negative consequences for their life quality. Politicians preached a litany of jobs and constantly rising living standards. What this effectively meant was mostly dreadful working conditions and the mass production of resource-guzzling rubbish, which was scarcely produced but had to be gotten rid of. Wellbeing and prosperity no longer consisted in the genuine satisfaction of basic needs, but in an addictive scrambling for short-lived consumer goods.

Christian festivals were emptied of their spiritual dimension and exploited as an excuse for more shopping. For children what was laying under the Christmas tree were dozens of plastic products imported from distant places. In the meantime they had lost the ability to sing or pray. In the course of the year every branch of industry got its turn, either through the exploitation of traditional festivals or by the institution of artificial, branch-specific festival days.

The historian, J. Wagenhals, designated this extremely destructive epoch as a consumerist pecuniocracy. Politicians never looked happier than at the cuttings of gaily-coloured ribbons to open new roads or other freshly paved surfaces. Then there were the times when they got together for group photos that appeared in the popular press. Each one acting like he is a public crusader brandishing a brand-new spade to "green" the asphalt desert by planting a tree. It was also characteristic for this era of capitalism-run-riot that it first took the protest of a generation of school children inspired by an underaged Scandinavian girl to bring the decisive change of heart. The result was for one thing that people were not only made aware of the imminent climate collapse but also took it seriously. For another, that the political elite's worship of production above all else was finally recognised for the insanity it was, and then treated with the contempt it deserved.

In Germany, however, the paradigm-shift initially went unnoticed, and when it was, the realisation came too late. The mass global market for cars reoriented itself. Not at least because people had become more discerning on the amount of extreme changes in the climate. High speed, suped-up horse-power and luxury fittings were no longer important. What people wanted was smart functionality and engine efficiency. Especially the latter took a long while finding acceptance in the aforementioned land of the car. There are countless reported instances of political functionaries speaking in favour of environ-menttally acceptable engine technology, but in practice they clung until the end to the noxious internal combustion engine. Even then it was already abundantly clear that if the calculations were done properly- taking into account the energy investment in the extraction, transporting and refining of the oil, the amount of heat generated in normal running and the actual fuel consumption needed to keep all those tons of car going- this kind of engine had an efficiency of under 5 %. Of every 100 litres of crude oil it used less than five to convey its body and payload along. The rest was dissipated into the atmosphere. At that time there were already other types of engine that performed very much better. Within a few years the market for petrol- and diesel- driven cars completely collapsed. Since the prosperity of aforementioned country had been based on them for decades, this created severe social problems. No one can say that they weren't warned. But they were the victims of a pathetically stupid business-as-usual attitude. The most ridiculous here were the so called ministers of transport of this era. They were often the least intelligent among the government ministers, and the fact that their efforts to promote a semblance of progress was actually just a drive to keep the destructive fossil-fuelled economy going gradually became clear to everybody. The inevitable crash from industrial prosperity to marginal poverty had never happened so fast anywhere in the world, except in countries involved in war. And those major wars, euphemistically called conflicts, were all about mineral resources, mostly battles over oil-reserves. They were cynically provoked by the rich countries, setting whole peoples against one another by exploiting religious differences in an incredibly perfidious way. It was the Black Age, named after the colour of oil and its black byproducts. It showed up everywhere: in the black of asphalt, in soot-blackened tyres and in poisonous exhaust fumes. What this actually referred to was the black oil being spread out over the earth like a deadly carpet and the air being saturated with it. Then the climate went crazy, and not until half the planet was suffering from thirst and the other half was flooding, did responsible politicians come to their senses. In many countries, as for instance in automobilian Germany, the turnaround only came after a general uprising. Right to the very last minute government ministers kept in place industry-friendly policies that today are inconceivable. On the one hand, with a stream of new decrees, apparently with the aim of improving air quality, people were forced to buy new cars at shorter and shorter intervals. Because they were at first considered environmentally friendly, these cars enjoyed tax benefits, but suddenly would be declared hostile to the climate and subsequently sold off to be used in other countries. As if climate wasn't a global process. On the other hand the interests of the manufacturers were always protected wherever possible. The waste of resources due to people being constantly forced to buy new cars was officially kept from public discussion.

It wasn't human intelligence that saved us, but pure necessity. Even during the period of rapacious capitalism there were already scientists and other individuals whose insight and integrity enabled them to prophesy that human action would soon bring the demise of spaceship earth. Yet the power- and prestige-hungry élite ignored every warning. At the final hour the leaders of the world's major countries were all non scrupulous liars and fraudsters. It were natural catastrophes resulting from climate change and life-threatening hardships, in other words widespread hunger and cold, that finally brought about the demise of this age of blindness. In many regions this assumed shape of extreme public unrest and revolutions. In what used to be called the First World the global collapse also let to a majority of well-informed citizens to reevaluate their privileges and finally to give up their crazy luxury lifestyles and by thus reducing their material needs. All of this let to a deflation of the economy. The whole over - heated, profit-maximising systems came crashing downright before they could entirely undo the planetary basis of human life.

Looking back from our present perspective, we can only be glad that this dark and destructive era of human history came to its end just in time for the earth's organism to be able to recover, albeit not fully.

She has arrived. The van stops directly in front of the place she had booked it for.

Although she is now in the city centre, the air is clear and pure and it is very quiet. All you can hear are the sounds of a few vans going past. By now everyone is used to how quiet the traffic is. There are nearly no more accidents involving pedestrians. In town the vans' maximum speed is 30kph, and they are equipped with emergency stop systems. Pedestrians for their part have become more alerted - they got used to things being quiet. Stories about the fossil-fuel era with its cities plagued with traffic noise and the stink of exhaust furies only make them shake their heads in pity and disbelief.

The other people taking part in the meeting have also arrived in a relaxed state, having been able to prepare themselves on the way. The meeting starts on time. The guests from the private company present their idea very convincingly.
Up to now, to get the payment on their account for the effort they have put into it, everyone has had to log into whatever MG they were using with their own personal energy card. MG is the abbreviation for mus-

cle-generator, a human-driven, stationary fitness machine which unlike those of the past, does not use but produces electricity. There are MG's in weather-proof halls, in the open air and in every long-distance train. They are activated by leg and/ or arm movements. Anyone can use them anytime for as long as they like. The electricity produced by this physical exercise is fed into the grid. For every 100 watt the person who produced them clocks up a fixed amount on his or her account. The motivation for using MG's varies. Many do it to increase their UBI. The UBI is paid out to all citizens every month. This unconditional basic income is enough to cover basic needs, but nothing more. Others, often high-earners with an interest in keeping fit use MG 's for sport. Many of them donate the money they make to charity. In that case the amount goes direct to the project they are supporting and as acknowledgement for this they are credited with an equivalent value in a virtual social currency. This currency has no material counterpart. It serves as motivation for altruistic action and

as an indicator of each individual's status in this regard.
The handles and seats of MG 's are provided with very thin, washable and 100% recyclable covers for the purposes of hygiene. There are, of course, still private fitness studios with everything of the absolute finest but they are subject to the luxury tax and so now they are few and far between. This tax is levied on everything except basic foods, clothes, toilet paper, soap and a few other exceptions such as every-day bicycles. Thus general consumption is kept down to a reasonable level. The tax revenues are used to finance measures intended to repair the damage done to the natural world and to sustain and further optimise the established cycles of materials, many of which are already closed.

The product presented at the meeting was a software update on the central MG server that leads to the possibility of future personal identity cards being read directly at the individual MG which would lead to an abolition of the need for an energy card.

She reported that this innovation has a lot going for it, since it enabled the energy card to be scrapped at very little expense and would also be perfectly in line with the basic principle of *less consumption means more value*. She promised the visitors that she would put their proposal before a higher authority and let them decide whether it should be presented to the executive committee. This committee- EC for short- has the ultimate say on the approval of publicly funded projects. It is composed out of four permanent and two floating members. Half of these are high-ranking politicians, the other half qualified scientists. Industry representtatives, economic functionaries, not to speak of professional lobbyists, who dominated such committees in the era of dark capitalism, are of course excluded nowadays. Back then the numbers on such bodies were always uneven, the business people having one seat more than the mostly more critical scientists. Thus all votes were pre-programmed in favour of industrial interests. Today the discussions are longer and the results correspond more to the demands of nature and the natural needs of the people.

The government now has a fundamentally different orientation. Whereas formerly ministers of trade were willing mouthpieces of uncontrolled industries, ministers of finance the vassals of greedy monetary institutions, ministers of transport asphalt junkies, environment ministers toothless lap-dogs and ministers of agriculture crafty trailblazers for an industrial agriculture inimical to animals and nature alike, nowadays we have ministries with powers running in exactly the opposite direction to all these. For example the minister of mobility is responsible for a transport system that conserves resources as much as possible, the flo-faun minister for the protection and conservation of plants and animals, the environment minister for maintaining the purity of air, water and soil. The latter runs the department known as Recyc-100 which oversees all manufacturing processes, making sure that every article produced can be 100% recycled, and that this actually happens. This department has far-reaching powers, extending to the suspension or permanent shut down of whole production chains.

In the afternoon she has an appointment with a representative from the Recyc-100 department. The complete makeover of the car industry contributed to the conservation of our planet's resources like no other measure. From our current perspective it sounds perverse that all the time when the extinction rate was at its peak constantly new "species" and varieties of cars were taking shape. At that time, via the internet, millions of individuals were able to put together "their" car - it was called "configuration" or worse "personalisation". The range of choices in terms of colours, fabrics or types of leather, engine capacity and extra features was gigantic and just as vast was the effort that went into production and spare-part management not to speak of the recycling of the highly valuable materials. Only the smallest fraction of all these materials landed up at the beginning of the production chain again. Today such excesses don't happen

anymore. Since throw-away capitalism severely diminished the stocks of raw materials, today it is essential that every substance is used for as long as possible.
The whole *eat-drink-and-be-merry-for-to-morrow-we-die* mentality of the past has given way to a more mindful, responsible attitude, thus geared towards sustainability.

Luckily there were already back then people who wanted to create and power cars as sustainable as possible. That's why mobile pioneers of a Munich start-up firm invented a minimalistic only including the necessary items, car in this otherwise misguided century, which only included the necessary items. On every side it was surrounded by solar cells, that's why it could be used for astonishing ranges back at the time. A later model of this for his time extremely futuristic car is now the only one allowed to be used in the individual traffic. Only thanks to the astonishing idealism of it's founders and the strong withstanding to gripping takeover offers made it possible for this project full of hope to not be eliminated by widely known producers and structures.

Though constantly up-dated, centralised design and manufacturing of all transport vehicles it has proved possible to build these with a minimal input of materials. Also the recycling of scrapped vehicles has been optimised to the highest level. The average life-span of the most widely used e-vehicle, the nine seater van, is two million kilometres. This has been made possible by regular visits to the municipal service centres and continual improvements in the quality of the replacement parts. With these vehicles everything is geared towards maximum efficiency of production, easy reparability and recycling and of course- in stark contrast to the way things were before - long life. These vans are in service almost all day and very very rarely break down. Noteworthy delays due to breakdowns do not occur as according to the size of the area being served, there will always be an appropriate number of vans on stand-by.

The Dept. of Recyc-Control has identified flaws in the mix of materials in seat-covers supplied by a private company, such that there will be problems with regard to their reusability. She was informed about this yesterday. This afternoon the official responsible for the case filled her in on the details. Immediately afterwards she contacted the supplier with a demand for an answering statement within 24 hours. If they do not succeed in meeting the requirement for 100% recycling capability within a set time period, the product will be subject to re-tendering. No compromises.

On her way home, with a few clicks, she books the journey for the winter holiday the family had chosen the previous evening: two adults, three children, outward on 23rd December, 2.20pm, returning on 2nd January at 11.15 am, cabin size medium, each way.
During the autumn the energy card function was transferred to the identity card. The private supplier didn't manage to solve the problem with the seat covers in the allotted time, so even before the winter holidays

arrived the contract was given to another firm.

On the 23rd December, a few minutes before departure time she and her family step into their comfortable cabin. Outside the temperature is 19 degrees. By turning a few handles they uncouple the cabin from the house. Scarcely have they shut the sliding door on the side where the cabin is usually joined to the house as an extra room and the children fetched their toys and games from the drover, when a slight jerk tells them they have been collected by the local e-transporter. This car is called to transport individual cabins over short distances. First of all they are taken to the local train station and there, together with other newly-arrived cabins, are loaded onto a middle-distance train without having to get out of their own cabin. After a few stops they arrive at the nearest mainline station. Here the children are amazed to see so many cabins and the parents are impressed at the speed and precision with which they are loaded onto their next train by the electro-cranes. Most travellers arrive on a pedelec, the majority taking theirs with them into the spacious bicycle carriage. There the batteries can be charged on the journey by high-energy chargers.

They don't have long to watch all this, however, for soon they are on their way on a long-distance fast-train. They'll be on this train for quite a few hours and so they soon feel the need to get up and move around. A side door takes them into the train corridor. Two tiers of cabins have been threaded together like pearls. They pass about fifty of them in all variations on the categories small, medium and large. With many you can look inside and see how different people's decorative tastes are. Then they come to the communal area, with its compartments and large open-seating areas. As they intend to cook something in their own cabin later, they resist the vegetarian enticements in the restaurant car. Her husband wants to take a detour to the bicycle carriage in search of tips for his next pedelec. To add to that there will also be some other travellers offering peds for sale. There are several rows of bikes, many equipped with partial weather shields. But there are no totally shielded models on board today. He has a chat with the owner of a USP- an ultra-speed ped. The owner says that on this bike he can do the 55 km to work in just under an hour in all kinds of weather and that when fully charged, the battery is good for 160 km. A super bike, no questions asked, but a bit over-the-top for his own personal usage. Another owner tells him about his LPD - long-distance ped. This bike has solar cells all over it, energy feedback from the brakes and a wind turbine that sets in on downhill drives, regaining energy. This leads to one battery charge yielding 300 km. What's the point, he wonders. After all, the network of rapid-charge points is really expensive.

As the light begins to fade they return to their cabin and soon there are steaming noodles on the table. "Look, mum! Autosaurs over there!". And sure enough there were a few of the noxious, smelly, fossil-driven metal crates limping along over dusty, potholed roads. Must be the racing season once again. For the people in the train this is a pitiful spectacle.

Nobody bothers to pay any more attention to this dying minority.

The cabin lights create a warm atmosphere for their family evening together in the cabin. Later they unfold the beds and scarcely have their heads hit the pillow, soon they are sound asleep. The parents wake up briefly as they are transferred a couple more times to other trains and then are unloaded at their final destination- the children continue being fast asleep. The next morning they are awakened by the sun. They have reached their destination, about 1000 km south of where they had started. They raise the blinds and enjoy the panoramic view of the mountains from their cabin. The parents explain that there used to be snow and ski-lifts here.
Being well rested, they open up the sliding wall and step through the short corridor to the main building.

It is a beautiful hotel with some traditional rooms for guests travelling without cabins. There is a wellness centre, and two restaurants. They enjoy every single minute of their holiday.

Wondering about the return trip? It will be even better because they will be travelling through the magnificent mountain landscape by day, which they slept through on the outward journey and with a little luck they might even see some frosty snow.
Why not this way? Why won't we chose differently?

Or another...

The sun glimmers palely through the swathes of acrid smoke. They reckon there must be water above them. They only allow themselves tiny sips of it. They wash with underground water only now and again, otherwise their skin burns. They're often cold so they burn the stinted bushes for warmth. They find lots of other stuff to burn too but then they end up with an unbearable acrid smoke cocktail. Many of the bushes have little sour berries. Game is rare but easy to hunt. They live in grey ruins, mostly close to the ground for these are less likely to collapse. The walls sweat brown water. Many of them have five or more limbs, some two heads. Rarely they have had children than they die themselves. Any who ventured away soon came back. They say things are the same everywhere. Some saw tall, round towers giving off a lot of heat. They soon ended up screaming and died horribly.

They all know that things on this planet were different for their ancestors; that they travelled by air all over the place and even more so on the ground, that they burrowed like beavers in the earth, digging mines and making a mess. At a relentless pace raw materials were turned into objects and thrown away as garbage just as fast, some of which is still lethal especially the stuff in the big round towers. They know that back then there were tall plants and large, powerful animals and that far more water was drinkable. They know a good deal more about their ancestors. Stuff that is of no use to them now. And they hate us!

Our last Warning?

Who do you think your are? You don't own sharp teeth or claws or warm fur. You are neither strong nor fast. You can do little in the water nothing in the air and were hunted on land. You shouldn't have stood a chance, normally. But then you took your only advantage, your ability to think, to be victorious over your enemies. You were the fittest to survive, you conquered this planet. And now you are in the middle of destroying it. You know this by heart but you don't change. From now on I will be with you at all times and l will hunt you. l am going to charge, again and again. So you can't extinguish me with your weapons. Those that you are going to create now as fast as the wind. You can't overcome me. You can put as much effort in it as you want. Many of you I have already killed and many more will follow. Like a beast of prey l will look for the weak ones, the old ones, the sick ones. I hope you will realise that you can't continue like this. Not here! There are worse ones in my family, severely worse. They are asleep in the deep woods hidden inside the animals there. If you don't change. They will also come to you. They won't make a difference between youth and old age or health and sickness. When this time will come you will live in fear. Like back then before you evolved to the leaders of this planet. That's why: change your mind, change your behaviour towards this earth!

Production and publishing:
BoD - Books on Demand, Norderstedt
ISBN: 9783755759201